D0604640

# Landslides, Slumps, & Creep

# Landslides, Slumps, & Creep

Peter Goodwin

A First Book

Franklin Watts A Division of Grolier Publishing
New York • London • Hong Kong • Sydney • Danbury, Connecticut

Photographs ©: Bob Barrett: 28, 29; Greg Gargan: 43; Peter Goodwin: 6, 37, 46, 48, 49, 53, 56; Photo Researchers: 19 (W. Bacon), 45 (Gregory G. Dimijian, M.D.), 23 (Ken M. Johns), 32, (Krafft - Explorer), cover (Tom McHugh), 47 (Jim Steinberg); Superstock, Inc.: 57; U.S. Geological Survey, Vancouver, WA: 30; U.S.G.S. Photographic Library, Denver, CO: 12, 13, 15, 22, 33; UPI/Corbis-Bettmann: 35, 39; Visuals Unlimited: 3 (Paul Bierman), 42 (Albert Copley), 11, 27 (John D. Cunningham), 20, 21 (Jeff Greenberg), 40 (David Matherly), 25 (James R. McCullagh), 51, 10, 17 (Steve McCutcheon), 16 (Martin G. Miller), 54 (Warren Stone).

Library of Congress Cataloging-in-Publication Data

Goodwin, Peter, 1951–
Landslides, slumps, and creep / by Peter Goodwin.
p. cm. — (A First book)
Includes bibliographical references (p. - ) and index.
Summary: Discusses the causes and consequences of landslides, avalanches, and other, sometimes rapid, sometimes slow, downward movements of rocks, soil, and/or snow.
ISBN 0-531-20332-8 (lib. bdg.) 0-531-15897-7 (pbk.)
1. Mass-wasting—Juvenile literature. [1. Landslides 2. Avalanches. 3. Soil mechanics] I. Title. II. Series..
QE598.3.G66 1997
551.3'07—dc21 96-37288
CIP
AC


Printed in the United States of America
1 2 3 4 5 6 7 8 9 10 R 06 05 04 03 02 01 00 99 98 97

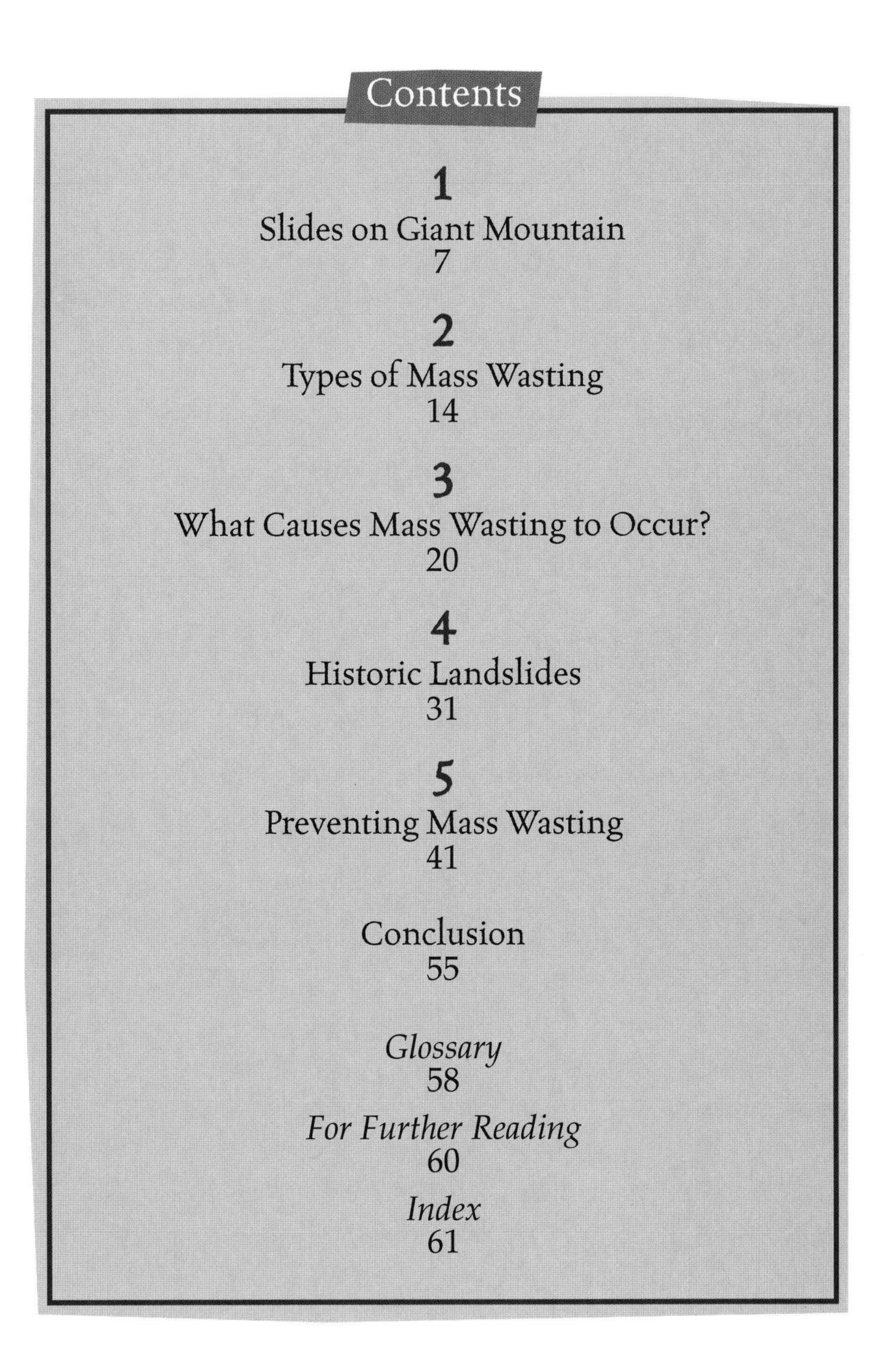

# Contents

# 1
# Slides On Giant Mountain

**June 29, 1962,** was a hot, muggy day in Keene Valley, a popular hiking spot in the Adirondack Mountains of New York State. A few thunderstorms passed through the area, but nothing seemed out of the ordinary. Then, one thunderstorm stalled over Giant Mountain. It rained and rained. Dirt clinging to the steep sides of the mountain soaked up the water and became increasingly heavy. Water soaking through the thin soils also made the solid rock under the soils slippery. Then, *gravity* started things sliding. The mountain shook as rock and dirt slid into the stream valleys.

The next morning, the tracks of fifteen major *landslides* scarred the face of the mountain. It would take two weeks to clear the debris off the main road below the mountain. A path of destruction 200 yards (183 m) wide and 3 miles (5 km) long followed the contours of a valley where a pleasant little stream had flowed through the woods.

◀ *The landslides of June 29, 1962, left their mark on the face of Giant Mountain.*

Debris sliding into the valley off the mountain's ridges caused the worst destruction. The sliding dirt and rock came to rest in streambeds, making temporary dams. As water rapidly rose against the sides of the new dams, the piles of debris gave way, unleashing a wall of mud, rock, and water that raced down the valley. It destroyed 200-year-old trees and sent them tumbling off Roaring Brook Falls, which drops 300 feet (91 m) in three steps. Water, trees, rocks the size of cars, and mud flew over the falls. The earth shook as they crashed into the ground.

The wall of water and debris then flowed onto a highway and washed away four cars. A boat, knocked off its trailer by the flood, floated down the road until it hit some trees. People camping at the base of the falls escaped unharmed but later found their tents sitting in 2 feet (60 cm) of mud.

Under clear skies the next morning, strips of rock freshly exposed by the slides glistened white in the sunlight, almost like snow. The familiar face of the mountain had changed. The slide paths soon had names that matched their shapes: the Bottle, the Question Mark, the Eagle, the Finger, and a slide that I named the Tulip. These slide paths, clean rock stripped of the vegetation that covered the rest of the mountain, became easy routes up Giant Mountain. Even the trail leading to the slide area was an easy hike. The wall of water had cleared the trees away and made it easy to walk up the stream to the slide paths.

In places, the streambed was 50 feet (15 m) lower than it had been before. In one place, the flood had cut a new streambed that took the water down a different valley, not over Roaring Brook Falls. Volunteers later blocked this new streambed so water would continue to flow over the beautiful falls. Although landslides can be much bigger than the ones witnessed at Giant Mountain, I was impressed when I saw them as a child. Tulip Slide still has the name I gave it.

## Mass Wasting

The landslides on Giant Mountain are examples of mass wasting. *Mass wasting* is the downward movement of rock and soil, often mixed with water. Sometimes the movement is rapid, traveling at speeds in excess of 100 miles per hour (161 km/hr). These landslides are dangerous because they catch people unaware.

Other types of mass wasting are *slump* and *creep*. A slump is the movement of a large amount of dirt and rock downward a short distance, building up a pile of dirt at the bottom that stops its motion. Creep occurs

*A slump moves a short distance and piles up at the bottom.*

very slowly, sometimes moving only ½ inch (1.3 cm) or less each year. If you come across a row of trees, fence posts, or gravestones that are all leaning downhill, it is probably a sign of creep. Another kind of mass wasting is a *snow avalanche.* In a snow avalanche, snow, often mixed with rocks and dirt, tumbles down the side of a mountain. Avalanches can be dangerous, especially to skiers who get in the way of the sliding snow.

▼*Creep causes these old gravestones to lean downhill.*

Some areas of the country experience mass wasting events very often. One such place is southern California. For most of the year this area is dry, but during the rainy season, water-soaked dirt often slides down the hillsides. Frequently, the moving mud travels rather slowly, giving people time to get out of the way. But these slow-moving slides bury objects in their path or knock them down. Buildings are often destroyed or sometimes just filled with wet, oozy mud.

Sometimes the California hillsides slump—dirt and rock quickly slide a short distance downhill. Other areas of the country, especially Alaska, experience creep when hillsides expand and contract as they freeze and thaw. Creep can break foundations of buildings, make railroad tracks become curvy, and break fences.

◀ *Even slow mudslides can damage a house.*

# 2 Types of Mass Wasting

**Scientists who study** mass wasting separate slides into many categories. A mass wasting event may be categorized by the type of material that slides, the shape of the slide, or the speed of the slide, among other factors. Some of the major categories of mass wasting include landslides, slumps, creep, and avalanches.

## Landslides

Landslides occur when rock, or dirt, or both start to move as a large mass. As a landslide slips over the material below it, there is a sharp division between the moving and nonmoving material, causing the slide path to have a distinct outline. As loose dirt and rock moved over the solid rock of Giant Mountain, the sliding material cleanly broke away from the nonmoving material. The different shapes left by the slides are reflected in their names.

Some landslides occur rapidly like the ones on Giant Mountain, but others occur more slowly. In some places in California, the motion is as small as a few feet a year. However, it is easy to distinguish the areas where there is motion from those where there is no motion. These slow-moving landslides can destroy houses just as easily as the faster kind. You have time, however, to get out of your house when the slow-moving kind occurs.

▼*A mudslide turns a dream home into a nightmare.*

## Slumps

Slumps are similar to landslides because there is a distinct line between moving and nonmoving parts. A slump, though, slides down a curved surface, and there is rotation of the material. When you "slump" in a chair, your head moves down and your knees move up. The rock and dirt in a slump move in much the same way. At the bottom of a slump, a large pile of dirt and rocks builds up and stops the movement of the debris.

▼*A slumping hillside in northwest Washington State*

## Creep

Creep is different from landslides or slumps because there is no observed line between the moving and non-moving parts, and it occurs extremely slowly. Creep can occur over large areas and is usually caused by freeze and thaw cycles or by wetting and drying cycles. Freezing of the ground causes the ground to expand because water trapped in the soil and rock expands upon freezing. This tends to break up solid rock and

▼*Creep resulting from freeze-thaw cycles on this Alaskan hillside causes trees to lean downhill.*

loosen the soil. Then, as the ice melts and the ground contracts, gravity pulls the loose dirt downhill. This process occurs in cold climates, especially in arctic areas such as northern Alaska. Wetting and drying cycles cause creep in a similar manner. If dirt contains clay, the clay expands when it becomes wet. When it dries, the loose dirt on top moves downhill.

## Avalanches

Avalanches are similar to landslides except that they are made of snow and ice with a little dirt and rock thrown in. If a layer of heavy wet snow falls on crusty snow, the heavy snow may start to slide. Freeze and thaw cycles also can increase the risk of avalanches. These cycles can cause some layers of snow to form larger crystals that act like ball bearings, allowing the top layers of snow to slide easily over them. At ski areas where there is danger from avalanches, specially trained personnel sometimes fire artillery shells at slopes that pose an avalanche danger. This causes controlled avalanches when there are no skiers to get in the way of the sliding snow. After the intentional avalanches, the area is probably safe.

▶*An avalanche on Mount McKinley, Alaska*

# 3 What Causes Mass Wasting to Occur?

**Mass wasting** occurs because gravity pulls everything downward. Forces in the earth cause mountains to rise, but mass wasting pulls them down. After the mass

wasting event, water and wind may remove the fallen rubble through *erosion.* The combination of mass wasting and erosion wears down mountains and eventually carries much of the material to the sea. Not all hillsides, however, are equally likely to slide. These are some of the factors that make a slope prone to mass wasting.

## Water

Water is often associated with creep, slumps, and landslides. This is because water affects the way dirt holds together. At the beach, sand castle builders know that dry sand doesn't stick together well. Damp sand, however, sticks together very well and can be made into vertical walls. If sand gets too wet, though, it just flows into a formless clump. In sand and dirt, there are tiny spaces between the particles. A little water in these spaces tends to hold the particles together with *surface tension.* When the spaces are completely filled, however, the particles are forced slightly apart and can flow.

On Giant Mountain, the heavy rains filled all the space between the dirt particles.

◀ *Builders of sand castles know that damp sand holds together better than dry sand.*

◀ *If too much water mixes with dirt, the dirt can flow downhill—possibly burying a populated area. Here, a man stands on a deep layer of dried mud after a mudslide has covered this street.*

This caused a landslide. In another case, a man in California left for vacation and forgot to turn off the lawn sprinkler. When he returned, his house had slid down a mountain, and the landslide had blocked a main highway. *Septic systems* can also add water to dirt and cause slides to occur.

Water can also cause solid rock to slide downhill. Water can weaken the *bond* that holds together layers of rock, or *strata.* Then, the rock layers start to slide just like dirt slides. The layers of rock are especially likely to slide if they are tilted downhill. If the layers are not tilted downhill, they are more stable and less likely to slide.

▶ *These layers of rock, or strata, tilt downhill. If the bond between them is weakened, top layers may break free and slide downhill.*

## Steepness

Along the coast of southern California, erosion by ocean waves is constantly making the oceanside slopes steeper. These steep slopes provide beautiful views and house sites, but as the slopes are made steeper, landslides and other mass wasting events become more and more common. Why is this so?

Imagine again that you are building a sand castle. If you take a handful of dry sand and let it fall into a pile, it will form a cone with a point at the top. The sides of the pile will have a certain steepness. As you continue to pour sand on top of the cone, the sides of the cone will get steeper and steeper. Eventually, the sides of the cone will give way, and sand at the top will slide down the sides of the cone and broaden its base. This will cause the sides to be less steep. If you pour even more sand on top after this has happened, the sand will build up at the top of the cone causing the sides to get steeper again. Each time the sides of the pile get too steep, a portion of the cone will collapse and slide. This cycle will continue as the pile grows, keeping the sides of the cone at about the same overall steepness.

Different materials will form piles with sides of different steepnesses. For example, damp sand will tend to form a cone with steeper sides than a cone made of dry sand. But whenever a slope is made steeper, it becomes more likely to slide. Mountain slopes are constantly being pushed up by forces within the earth, and rivers and oceans are constantly eroding the bases of hillsides. These forces are constantly creating steeper slopes, and gravity is constantly trying to pull these slopes down.

▼*As the ocean erodes the base of this steep California shoreline, the cliffs become steeper and mass wasting is more likely to occur. As this cycle continues, the edge moves precariously close to buildings at the top of the cliff.*

## Climate

Climate also plays a big roll in causing landslides. In California, most of the year is very dry, and wildfires and drought sometimes kill the vegetation on hillsides. When the vegetation is killed, the roots die and can no longer hold the dirt together. Three or four years after a fire, the loose dirt may be ready to move downhill. Winter storms sometimes dump lots of rain very quickly, and the water mixes with the loose soil and makes oozy mud that flows down hillsides and into streams.

## People

People can increase the chance of mass wasting by altering the landscape during construction. Often, people cut into slopes to build roads or houses. By steepening the slope, a landslide can occur more easily. Likewise, if extra dirt, or fill, is added to a slope to make a flat space to build a house, then the extra weight can bring the whole hillside down. Also, people may divert streams or add septic systems that increase the amount of water in the ground, making it more likely to give way. Or, they may clear a hillside of vegetation, depriving the slope of roots necessary to hold the soil in place.

▲*Fire has stripped these California hillsides of live vegetation. As the roots die and can no longer hold the soil in place, the hills become prime targets for mass wasting.*

If mass wasting is common in a proposed construction site, then engineers must plan very carefully to avoid contributing to slides.

Virgin River Canyon
Recreation Area
NEXT RIGHT
CAMPING

*When construction, such as this roadcut, makes slopes steeper, they become more prone to mass wasting.*

# 4 Historic Landslides

**Many dramatic landslides** have occurred in recent history. Most of them cost people their lives. They have occurred all over the earth—the most spectacular ones occurring on steep slopes. Steep slopes cause dirt, rocks, and perhaps snow to slide so quickly that people below have no time to get away.

## Mount Saint Helens Landslide

One of the most famous landslides in the United States occurred on the slopes of Mount Saint Helens in Washington State. In 1980, the volcano erupted. The explosion blew down millions of trees on the slopes of the mountain and killed a number of people. Landslides carried the destruction far from the mountain.

◀ *The eruption of Mount Saint Helens melts snow and ice on the mountainside, which mixes with dirt and rock and flows downhill as a massive landslide.*

Lava and hot gases rocketing from the volcano melted huge amounts of snow and ice. The water mixed with dirt and volcanic ash on the slopes and raced down the mountain. The dirt and mud filled the streams around the base of the mountain and wrecked everything in its path. Miles away, mud moved a highway bridge hundreds of feet from where it was supposed to be.

▼*Houses caught in the Mount Saint Helens landslide are easily swept up in the advancing mud.*

## Gros Ventre Landslide

In the spring of 1925, a large landslide occurred in Wyoming. Huge amounts of dirt and rock slid off a mountain and dammed the Gros Ventre River. The slide occurred because rock strata tilted down toward the valley at an angle of about 20 degrees. The weak bond between the layers of rock couldn't support the weight of rain-soaked snow and dirt. Rock and dirt slid down

▼*The mountainside in the background still shows scars from the huge 1925 landslide that dammed the Gros Ventre River.*

the slope into the valley below and formed a dam about 225 feet (70 m) tall. A 5-mile-long (8-kilometer-long) lake formed behind the dam.

A year and a half later, the dam broke because the loose dirt eroded quickly as water flowed over the top of the dam. The flood killed six people in the town of Kelly, a few miles below the dam. Some people were standing on a bridge, watching the flood waters rise. The flood destroyed the bridge they were standing on. The whole town of Kelly, Wyoming, was destroyed, but it was later rebuilt.

## Ranrahirca-Yungay Landslide

In Peru in 1970, a landslide destroyed two towns, Yungay and Ranrahirca. Two years before, American hikers climbing Mount Huascarán had observed a huge block of ice on a *glacier* that seemed ready to fall. They warned the government, and people were worried for awhile but then forgot about it. Then, an earthquake shook the mountain, and the ice block broke free. It started down the mountain bringing lots of rock and dirt with it. The fall caused the ice to melt. The water mixed with dirt and rock to form a muddy mass that rushed down the mountain.

▲*This mangled bus in Yungay, Peru, demonstrates the power of the 1970 landslide.*

The slide fell 12,000 feet (3,658 m) through the air before crashing into the side of the mountain and continuing into the valley. In all, the slide traveled 9 miles (15 km) in less than 4 minutes at an average speed of about 135 miles per hour (217 km/hr). Some of the slide buried the town of Ranrahirca, killing 1,800 people. A small part of the slide went over a ridge and flew through the air. It landed on the town of Yungay, killing 17,000 people. Only the church steeple stuck up above the mud.

The few survivors waded through the mud trying to rescue others, but they couldn't really move through the area for about three days. By then the mud had dried enough to walk on. Today, flowers and small trees grow on the site of the slide. The place looks peaceful. Scientists who have studied the area say that the slide of 1970 was not the first one. Other slides had come down the same track.

## Crawford Notch Landslide

A landslide occurred in New Hampshire early in the settling of the White Mountains. It occurred in 1826 in Crawford Notch. Some early settlers, the Willys, had built their home and farm buildings below a steep mountain slope. The remains of a tropical storm dumped huge amounts of rain on the steep hillsides above their home. Although no one knows exactly what happened, there must have been a huge roar as rock and dirt raced toward the house. Everyone must have panicked and run, only to get caught in the landslide. The landslide split in two just above the house, leaving it completely undamaged. Ironically, had the Willys stayed in their house, they would have escaped unharmed. Two oxen survived because they couldn't move from their stalls.

▼*The innocent-looking clearing on this mountainside is the remnant of an 1826 landslide that buried the Willy family.*

## Vaiont Dam Landslide

In Italy, on October 9, 1963, a landslide slid into the lake formed by the Vaiont Dam. The slide occurred because rock strata pointed down toward the lake. Rain from many storms had saturated the dirt and added weight. The slide carried more than 262 million cubic yards (200 million cubic meters) of rock and dirt into the lake behind the dam. This caused a huge wave that splashed over the dam creating a wall of water 328 feet (100 m) high. This water flooded the valley below the dam and killed almost 3,000 people. The strongly built dam did not break. Had it broken, many more people would have died.

## Other Slides

In 1959, an earthquake in Montana caused the side of a ridge to fall into a river valley. The slide dammed the river and formed what was later named Earthquake Lake. In the Adirondacks of New York, a slide in 1942 came down the side of Mount Colden. The slide poured into Avalanche Lake, raising the water level about 9 feet (3 m). Bridges had to be built around the edges to allow hikers to walk where they had once walked on dry land.

▲*The aftermath of the landslide-induced flood beneath Vaiont Dam*

▲*A view of Earthquake Lake, created by a landslide off the mountainside in the background. The dead trees in the lake were drowned as water backed up behind the slide.*

The list of landslides goes on and on. History shows that where there are steep hills, earth and rock will eventually slide. It can be extremely difficult, however, to pinpoint exactly when and where slides will occur. Methods for predicting and preventing slides are constantly being improved.

# 5 Preventing Mass Wasting

**If you are building** a house, road, or dam in an area prone to landslides, slumps, and creep, you must think about what can be done to prevent mass wasting. Otherwise your construction may be destroyed and people may be injured. People who own property already affected by landslides, slumps, or creep look for ways of preventing further damage to their property. But what can be done to prevent gravity from pulling soil and rock downward?

The easiest way to avoid the dangers of mass wasting is to build on level surfaces. This, of course, is not always possible. Roads must go through mountain passes and along streams with steep banks. Houses on hills are desirable because they have views that are not possible from level surfaces. Therefore, engineers design the construction to reduce the dangers to the smallest levels possible.

▲*An avalanche track*

## Identifying Risky Areas

Often, loss of property through mass wasting can be avoided by looking at the history of the area. In some steep mountainous areas, scientists can simply look for old landslide tracks and assume that slides will eventually happen there again. Avalanches often follow the same tracks year after year because they get funneled into the same valleys. In these areas, trees don't grow because they keep being removed by the avalanches.

If a road must pass through an avalanche area, building a roof over the road may allow an avalanche to go harmlessly over the road.

Not all mass wasting leaves traces as obvious as the slide tracks left by avalanches and major landslides. Often, *geologists* look under the surface of the land. They might take a *core sample* and analyze the soil types. They may be able to detect subtle downhill movement of the soil.

A core sample is taken by drilling down into the ground with a hollow drill. The hollow drill pulls a

▼*This shelter protects the road from frequent avalanches.*

column of dirt out of the ground in the same way that an apple core is taken out of an apple. This "core" is then examined to determine the soil types and to see if there has been any slippage in the past. Soils that may cause problems include those with high clay content, because clays tend to become slippery when wet.

Geologists also look at the rock layers, or strata, when determining the chances that an area will slide. Because it is easier to slide down a ramp, geologists look for rock layers that point downhill. Then, they look at the type of rock. If the bond that holds the layers together is strong, then the rocks may never slide. Lots of mountains have sloping strata and don't slide. However, if the geologists detect weak bonds, then they start to worry. There are many techniques, however, that geologists can use to make an area less prone to mass wasting.

## Techniques for Preventing Mass Wasting

Road builders work to make sure that slopes don't slide by being careful about how steep the slopes are. They take extra precautions if the rock strata tilt downward toward the road. If there is danger of sliding, sometimes the top of a hill is removed, or "steps" are built into the hillside to reduce the chance of a slide. Removing the hilltop takes away extra weight that might cause slipping.

Along steep slopes beside highways, you often see fist-sized rocks covering the hillsides. These are placed there because they allow water to flow easily down the slope while holding the soil in place. At times, engineers use *gabions*—wire crates that are filled with rock. These act like big blocks of rock and do not slide. Cement walls are sometimes used instead of gabions.

▼*Gabions on a hillside*

In some areas, individual rocks often fall off cliffs or hillsides onto a road below. To prevent this, crews may drill holes in the rock and place large bolts in the holes to hold the layers of rock together. If you look at rock cuts along highways, you can sometimes see these

▼*These rocks are held in place with large bolts.*

▲*Wire fences keep rocks from sliding onto the road.*

rods pointing out of the rocks. *Rockfall* can also be prevented from interfering with traffic by building steps into the slope beside the roadcut. Rocks fall onto the steps but do not roll onto the road. In other cases, wire fences are constructed to literally catch the rocks and prevent them from landing on the road. All of these systems prevent disruptive rockfall most of the time. However, some rocks, especially big ones, will still end up on the road. Remember, these systems reduce the danger, but they don't get rid of it.

*The bolt sticking out of this fallen rock failed to hold it in place. Luckily, it did not fall on the road.*

When building houses on steep slopes, engineers try to keep the slopes as gentle as possible. When the slope is steep, a level place must be made for the house. Engineers try to create this level space by cutting back into the hillside and removing excess dirt and rock. The alternative, creating a level space by adding dirt, called fill, is usually avoided. This method adds weight to the slope and makes the slope steeper beneath the house; both of these factors can make sliding more likely.

Engineers also try to avoid adding water to the soil. They design septic systems so that they drain water to places where the water won't cause sliding. Drains can also be placed in the hillside to remove ground water. By reducing the amount of water in the ground, the land becomes more stable. In extreme cases, engineers install pumps to actively draw water out of the ground.

Damage from mud slides can sometimes be prevented by building pits that can contain sliding mud, keeping the mud from reaching populated areas. These "mud ponds" are built in valleys that tend to have mud slides. After a mud slide occurs and the mud dries, the dirt is removed to make ready for the next mud slide down the slope. These mud ponds work as long as the slides are not so large that they fill the pond and then continue down the valley.

Creep is sometimes a problem around houses, even

those built on relatively gradual slopes. The cause of creep is usually related to soil type. Clay soils often promote creep, especially if other dirt has been placed on top so that a lawn can be grown. Creep can be prevented, or stopped, by placing steplike terraces on the hill. The foundations for the terraces must be dug down into the subsoil, the soil below the surface. The terraces

▼*These terraces are designed to halt mass wasting on this Alaskan hillside.*

can convert an entire slope into a series of level surfaces. If they are built correctly, these terraces can prevent creep. If they are not well built, the terraces can creep downhill, too.

## Preventing Mass Wasting Due to Erosion

Geologists worry about mass wasting in areas where slopes are being made steeper for one reason or another. This often happens where there is moving water. As streams wind back and forth in their valleys, or as ocean waves hit a steep shoreline, the water cuts into the bordering slope. These slopes get progressively steeper and may eventually collapse into the water. The moving water then carries the fallen material away and the cycle continues.

As shores become steeper, the chance of mass wasting increases. When construction has to take place near streams or the ocean, engineers often try to stabilize the shore by large placing rocks, called *rip-rap* along it. This is often done near bridges. The large rocks will not

▶*As the waves continue to make this shoreline steeper, mass wasting will constantly occur. This may cause problems in populated areas.*

erode easily, and they prevent streams from eroding bridge foundations. They also prevent the steepening of river banks and ocean shorelines.

▼*Rip-rap controls the erosion of this bank.*

# Conclusion

**Mass wasting** will always occur as long as the forces in the Earth keep raising up land to make hills and mountains. In places where the mountains have been rapidly raised and where slopes are steep, landslides will occasionally occur. Even on gentle slopes, gravity tends to move dirt and rock downhill.

The downward motion may be gradual and just cause gravestones in a cemetery to tilt downhill after a hundred years. Or, the motion may be a dramatic landslide in which the entire side of a mountain collapses. In all cases, gravity causes the sliding. Over time, old mountains will be worn down to sea level, and mass wasting will cause much of the movement. The rest will occur as rivers and streams carry the dirt and rock away. But as this happens, new mountains will rise. The face of the earth is constantly changing, and mass wasting is an important part of this transformation.

▲*Mass wasting is constantly occurring on these hillsides, making them unsuitable places to build homes.*

People can live in an area where mass wasting occurs, but precautions must be taken. With care, the chances of having your house slide down a steep hillside can be small. Likewise, if you are careful, you can build in a place where landslides won't fall on your house from above and destroy it. In some cases, slides can be slowed or stopped by doing such things as removing water from the ground or preventing further erosion by streams. Obviously, mass wasting events can be dangerous. However, with careful planning, we can live safely where the ground slopes. Proper planning can keep a home safe for a long time.

▶*With careful planning, hillsides can be beautiful and safe settings for homes.*

## Glossary

*bond*—the natural force that holds substances together.

*core sample*—cylindrical section of earth or rock removed for study.

*creep*—slow downhill motion of soils.

*erosion*—the process of wind and water wearing away soil and rock.

*gabions*—rocks placed in wire cages placed on hillsides to prevent mass wasting.

*geologist*—a scientist who studies the history of the earth, especially as recorded in rocks.

*glacier*—a very large, slow-moving mass of ice.

*gravity*—the force that pulls objects toward the center of the earth.

*landslides*—downhill motion of earth and rock with definite areas of sliding and nonsliding material.

*mass wasting*—any motion due to gravity of rock, dirt, or snow downhill.

*rip-rap*—large rocks placed along shores to prevent erosion.

*rockfall*—mass wasting of distinct pieces of rock.

*septic systems*—underground spaces used to store waste water from houses.

*snow avalanche*—the sliding of snow or ice down a mountainside.

*slump*—motion of a section of dirt and rock for a short distance downhill, piling up at the bottom.

*strata*—layers of rock.

*surface tension*—the property of the surface of a liquid to pull together.

# For Further Reading

Allen, Missy and Michel Peissel. *Dangerous Natural Phenomena.* New York: Chelsea House, 1993.

Goodman, Billy. *Natural Wonders and Disasters.* Boston: Little, Brown, 1991.

Lauber, Patricia. *Volcano: The Eruption and Healing of Mount St. Helens.* New York: Bradbury Press, 1986.

Van Rose, Susanna. *Earth.* New York: Dorling Kindersley, 1994.

# Index

# About the Author

**Peter Goodwin** has taught at the Kent School in Connecticut for more than twenty years and is Chairman of the Science Department. He teaches a course in geology as well as courses in physics, astronomy, and meteorology. He enjoys the out-of-doors and finds that geology can add to the outdoor experience. Mr. Goodwin has written many books on science projects and physics. He has a degree in physics from Middlebury College and a degree in education from Trinity College.